INSTANT CHRONICLES

D. J. ENRIGHT

INSTANT CHRONICLES
A Life

Oxford   New York
OXFORD UNIVERSITY PRESS
1985

Oxford University Press, Walton Street, Oxford OX2 6DP
London New York Toronto
Delhi Bombay Calcutta Madras Karachi
Kuala Lumpur Singapore Hong Kong Tokyo
Nairobi Dar es Salaam Cape Town
Melbourne Auckland
and associated companies in
Beirut Berlin Ibadan Mexico City Nicosia

Oxford is a trade mark of Oxford University Press

© D. J. Enright 1985

First published 1985 by Oxford University Press

All rights reserved. No part of this publication may be reproduced,
stored in a retrieval system, or transmitted, in any form or by any means,
electronic, mechanical, photocopying, recording, or otherwise, without
the prior permission of Oxford University Press

This book is sold subject to the condition that it shall not, by way
of trade or otherwise, be lent, re-sold, hired out or otherwise circulated
without the publisher's prior consent in any form of binding or cover
other than that in which it is published and without a similar condition
including this condition being imposed on the subsequent purchaser

British Library Cataloguing in Publication Data
Enright, D.J.
Instant chronicles: a life.
I. Title
821'.914 PR6009.N6
ISBN 0-19-211971-0

Set by Getset (BTS) Ltd.
Printed in Great Britain by
J.W. Arrowsmith Ltd.
Bristol

In biography, the fiction parts should be printed in red ink, the fact parts in black ink.
Publishers Weekly

There is no life that can be recaptured wholly; as it was. Which is to say that all biography is ultimately fiction. What does that tell you about the nature of life, and does one really want to know?
Bernard Malamud, *Dubin's Lives*

Lives of the poets

'They are present in the radiance of their works . . .
Like lights that blaze so bountifully on summer nights
   from bowers and lawns,
Like stars on earth, or diamonds, emeralds and rubies
Left in the bushes by an emperor's children at play,
Or raindrops hung in the high grass, sparkling with joy . . .

O never let us scan their lives at close quarters!
Forbear to seek those stars, those jewels, those raindrops
   in the cruel light of day—
All we shall find is a poor discoloured worm, as it crawls
   through the mud.
The very sight disgusts us. Only a nameless pity
Stays us from crushing it underfoot . . .'

This from Heine, who must have had his reasons. And of Shakespeare, no less. As worms go, not the worst of creeps. But in general the cap fits. Distinctly romantic in manner, of course. Stormy and drangling, like Hatton Garden on a wet day. What can he mean by nameless pity? It's a pitiful biographer who fails to come up with a name. Close quarters . . . and no quarter! Perpetual daylight is our cry.

Prefaces

Was that the primal scene?
It should have been.
And yet he was so young then.
Scene, obscene, unseen?

How could he see a scene?
He had no eyes then.
Nine months to go. But even
So, all the best babes have seen.

> Would he see the final scene?
> Let it not be obscene.
> When what must be would be,
> Not much to see.

Growing up lonely, a tadpole against the tide,
He could never put his trust in mass emotion.
Such was his horror of the mob, or else his pride,
That if he noticed people thronging into heaven
With drums and trumpets and Peter as their guide,
He would run at once in the other direction.

> As the twig is bent
> So is the tree inclined
> Bent so often, so many inclinations
> Could hardly see the tree for the woulds
> The murky mights, the tiny maybes
> A laurel bough endures the usual charring
> Branches get cut that sometimes grew.

While waiting for a ship to Egypt—
Filling in at a private establishment
For little misfits from the middle classes,
The slightly backward, the rather delicate.
They seemed to him nice ordinary kiddies.

Mrs Headmaster had her personal family.
He often met her at the foot of the stair,
A stately matron bearing a brimming chamber:
'Milk and honey!' she pealed out proudly.
There was a power of rude health up there.

Noises came through the gilded ceiling
Of furniture breaking, of angers and hungers,
The cublike romping of upper-class infants—
While he and his tender flock sat trembling.

Great debate

Swift lamented that Celia, Celia, Celia – – – – – [rhyming with *spits*], but along came Lawrence, crying: ridiculous! monstrous! of course Celia [rhymes with *spits*], and just as well, it being her proper natural function! And so battle was joined. To *spit* or not to *spit* was the burning question. (Not merely when or where.) Some said: great! beautiful! the more *spitting* the merrier, it was great Nature! Others felt: but Nature could have been nicer and spared the already humbled from crouching even lower. It was the hearties and the aesthetes, the physicals and the metaphysicals, all over again, but more so. This was the great debate. It still rages, though not always stated in these particular terms.

Explanations

His folk were excessively averse to exaggeration
And other extremes. They said what they meant.
(Which might imply a poverty of ideation.)
In their opinion, tragedy was best evaded;
They lacked the resources. Comedy had its place,
When it wasn't out of it. Mere misery degraded—
By semantic shift they talked about the weather.

Hardly an auspicious kick-off for an author!
It explains a lot, not least his exaggerated fear
Of exaggeration. And an unprofessional *pudeur*.
(He was rarely heard to talk about the weather.)
His people favoured short sentences and the wisdom
Of simple facts. If depths there really were,
Depths weren't put there for them to fathom.

No wonder that foreigners, the most outlandish,
Were the only subject he felt at home with—
As if they'd want a literature, and one in English!

Absit omen

In Egypt, his little world reeling,
He was sitting and typing his thesis,
'The Search for God'. His skin was peeling.
Scratch, scratch. Drifting away in pieces.

He noticed the flakes were vanishing,
Slowly but surely under the floorboards!
The search was over. God was ravishing
Him shred by shred, dragging him downwards.

Ants it was, bearing away every feather
Of skin for nests, for nutrition.
Symbols were drifting about him, as ever—
A circumstance he was loath to mention.

Catch

Slow to destroy life whatever form it shows
Except for cockroaches, fleas and mosquitoes—
We know the hell reserved for such as these:
Beset by mosquitoes, cockroaches and fleas.

Stones

When a taxi-driver threatens to summon
The Moral Police, that new arm of the law,
And accuse you of criminal conversation
In the back with an innocent Egyptian whore—
Then, albeit at the time you were quite alone,
Be prompt to pay him whatever he asks for.
Foreigners shouldn't cast the first stone.

. . .

There were sermons in stone,
He remembered, as one whizzed past his nose,
Just missing the policeman who had nabbed him.
The populace were on their toes:
Foreigners were spies, as well as improper,
And some were Jews.
The policeman soon decided to let him go—
Perhaps after all his papers were in order
(Being illiterate, how could the man know?),
Loitering though he was, or moving furtively fast,
In the vicinity of a blatant anti-aircraft gun.
Anybody can get hit when stones are being cast.
You can never find a taxi when you want one.

Appearances

Like high priests disguised as shepherds, there they were, tall figures in immaculate robes, bearing staffs

They sat in the harbour bars, undrinking, unspeaking, like pharaohs disguised as high priests

Wherever he went they were there, imposing, impassive, like archangels disguised as pharaohs

Coffees were hastily ordered, a hush fell, they sat there in state, like gods disguised as archangels

A friendly drunk avouched in a whisper: their staffs were loaded, they were secret police in disguise

The bars were sweetness and light, as there they tarried, like hefty humans with knobsticks disguised as gods

Till they decided at last that he wasn't a Zionist agent disguised as a teacher of English

And they faded away from men's eyes, like guardian angels disguised as optical illusions

Then petty crime revived in the bars, disguise was discarded, unfriendly drunks threatened to knife him.

Innocence

Days of innocence! A knob of hashish nestling
In a mound of damp tobacco. An urchin
Running with relays of red-hot charcoal.
What giggles! Good for the lungs it was,
Sucking at the hookah; if you had good lungs.
They passed the tube around, from mouth to mouth,
Difference of race or caste no more they knew.

Grass was a green stuff then, in short supply.
Cannabis? Something to do with the old embalmers?
Hemp you wouldn't mention in the hanged man's house.
Marijuana could only be a daring Spanish dancer.
(Just to kiss a girl you'd have to marry her.)

Once a windfall: broken biscuits of the stuff
Found in the bottom of a boat that plied the Nile;
No doubt the boatman had been smuggling it.
Then as the sun sank in its wonted splendour
How high they got! Later it proved to be caked mud
Dropped from the boatman's toes. Fine hashshashin!

And now? One at least worn out or dead in gaols;
A fat effendi sweating over files; a cotton broker;
A colonel long cashiered; a weary village teacher;
The clever one in Paris, still a starving artist;
And one remembering those days of innocence.

Question of degree

So they fled from Egypt, with the child in the womb, who was safely delivered in Rowley Regis, not far from Birmingham University. Soon after there came HM Inspector, asking about his doctorate. An Egyptian degree? Was it recognized by the Ministry? He didn't know? He hadn't bothered to find out? Tut tut! But no more was heard of this delicate question. The Ministry didn't know either.

Circumstance

Adult classes in Cradley Heath were full of characters;
They had been together for ages, they never grew tired of themselves.
Among them, a retired schoolmarm, a bit of a Buddhist,
Who doted on poetry and found the others unrefined;
An old-time chainmaker referring everything to Marx;
A knowing little grocer spotting sex behind each scene
(Such disappointment when they came to *Sons and Lovers*);
A black-browed preacher who allowed of only one good Book;
A motherly housewife representing wives and mothers.
You always knew where they stood, they all had circumstance pinned
   to the ground.

He would think of his other students, in Egypt—
Sent down to prison for rioting out of season,
Or summoned back to villages to feed their families.
Several went crazy, one of them during a lecture:
Was he berating Britain? No, the others assured him,
No need to worry, he was only thinking he was Allah.
A few rich boys waiting to inherit and abscond to Paris;
And the shy bright girls, bunched together in the front,
Touching their cheeks to check they'd left their veils off.
You never knew where they were, circumstance had them all by the
   throat.

Animal lover

      Once, something tells them, it was all theirs.
      A fair amount still is.

      The burden lifted from them,
      These days they pledge themselves to leisure,
      Cultivate their tails in others' gardens—
      Quondam statesmen, finicky about food,
      Twitching their whiskers at birds in bushes.

Poetry shines from behind their eyes,
But they will not read it to us,
Content that we know what they think of us
(And still admire them),
Famous for certain though unspecific wisdom.

Their successors have failed to rid the world
Of dogs; and even regard them as friends.
Ah well, that regime will soon be over too;
Surprising they ever got themselves elected.
They too will turn into predecessors—

And remembered for some mysterious wisdom?
A cat yawns sceptically: Wouldn't bet on it.

The pattern was set by a young tomcat called Tom who turned up at the door one evening. Tom was a fighter, and by the looks of it Tom was a loser. Tom never won fair lady. He developed abscesses and sat gloomily on the mat while they dribbled like tired volcanoes. He was love's victim.

One way to stop this, said the vet as he reached for his scissors. Poor Tom would nevermore desire to win fair lady. He put on weight and muscle. His motives were unmixed, his claws pure, it was the Way of the Samurai. He tore his old rivals apart—abscesses were for others—and chased off the neighbouring boxer, slash! across its broken nose. He gained the respect of all. Fair ladies fell at his feet and were spurned. Tom became a power in the alleys. His motto was: With our stripes are we healed.

First impressions

At Tokyo airport, descended from a Comet,
He is asked what he thinks of Japan.
The only thing he can think is—Japan
Is agog to know what he thinks of it.

These days he wouldn't dare to ask
What they might think of Britain.
And these days would they bother to ask
What he thought, if he thought, of Japan?

Going cheap

There are some will be heard on Judgement Day
Wailing that their souls have been underpriced.

Invited to do a little informal informing
On the politics of Japanese intellectuals
('To you they'll talk freely'),
What stuck in the throat was the rate for the job.

An occasional trip to the great metropolis
With bed and breakfast at a three-star hotel—
A real treat! All paid for by the Embassy.

Do not litter

A day trip to the sacred island
Where birth and death are both illegal—
Men throwing out their chests and hooting lustily,
Women pulling stomachs in and looking virginal,
Boats on the beaches waiting eagerly
To ferry suspects to the mainland.
Hiroshima is the nearest city.

No show

Stripped to the buff the Japanese girls trip on—
One of them wrestling with a torpid python—
Slowly slowly *dan-dan* they put their clothes on—
Even the python stirs—oh the mounting tension!

Entertaining women

In a night-club in Hiroshima,
A combo playing noisily,
A girl asked sweetly, *Kohi shimaska?*:
Should they make coffee?
No, he replied, it kept him awake.
It was *koi*, it struck him later, not *kohi*:
It was love she had offered, not coffee.
The thought kept him awake.

Next day, as a guest of Rotary,
He conveyed (without authority)
Fraternal greetings from Cradley Heath.
Waiting outside was a victim
(Rotary does not entertain women),
A victim for him to see, to see him.
Him with his face still scarlet,
Her with her white scarred arms.

A child's view

The amah was Mrs Okamoto, a good Christian
And her loving crony. How proud the child was
When Christmas came and her dear one's praises
Were sung by all: 'Okamoto ye faithful!'

They pored over pictures of Hiroshima: ashes,
Shadows of people printed on a bridge, a hole
Burnt in a youngster's pants ('It must have hurt!').
Who was it did those bad unchristian things?—
The Aremicans it was, the cruel Aremicans.

There were foreigners scattered here and there—
A few Eikokujin and the indistinguishable
Amerikajin, and the more exotic Furansujin
Like her mother, some Orandajin, an old Doitsujin
Teaching German at her father's college.

But none, not one, of those heathen Aremicans
Who burnt holes in you and turned you into shadows.
Just thinking made you shiver. But not too much.
They were far away, you would never meet them.

Refinements

'There's a lot of loose talk about the *feudal society* of the bad old days,' complained the Prime Minister's brilliant son. 'Also about the miserable condition of women. Let me tell you a story.' And he did.

Of how in those old days a gentlewoman, travelling in palanquin or carriage, might experience certain known though unadvertised urges. Whereupon, feeling faint as is a lady's prerogative, she would halt at a country inn and ask to lie down for a while. Bowing low, the innkeeper would reply that his wretched establishment was indeed honoured: pray grant him a few moments to prepare a room.

She lay on a thin pallet, taking her ease. Underneath (perhaps less at ease) lay a sturdy village lad, speedily procured and primed. Neither took note of the other, despite the curious apertures in the pallet. What is not observed is not official, what is not spoken of has not happened. Eventually she would rise, feeling refreshed, and thank the innkeeper for his hospitality, while pressing into his reluctant hand a number of coins. Some of which, we trust, found their way to the village lad, by now back in the fields.

'Such refinements are the mark of a superior civilization,' the Prime Minister's son declared. 'So let's hear no more about that stuffy and repressive *feudal society.*'

Samurai

Those squat fellows, like worried gangsters—
Was it the presents they were guarding?
No, they were keeping out the bridegroom's
Army of strange friends and queer admirers.
You wouldn't want too many at a famous wedding.

Aesthete or soldier, it's sincerity that matters.
Later he sought to raise a Patriotic Army,
But the troops preferred a quiet life.
They didn't approve of imaginative writers
Shouting out eccentric orders.

So he slit his belly, to show them,
Crying at the last, 'Long live the Emperor!'
Something they would rather not have seen.
At that time such courage was in error.
How embarrassed the Emperor must have been!

At the theatre

With his faithful follower Benkei and a few retainers, Yoshitsune approaches the barrier at Ataka. It is guarded by soldiers of his jealous brother, the Shōgun, who is hounding him. The fugitives are disguised as mendicant monks, except for Yoshitsune, who is clad as a coolie. 'If you are monks,' says Togashi, the captain of the barrier, 'you must be collecting for some temple. Pray read out the subscription list.' Benkei pulls out a paper and pretends to read from it. Togashi sees through the ruse. He is a man of honour. So are they all. Togashi makes a donation, and allows them to proceed. One of his guards recognizes Yoshitsune: 'Stop!' Benkei hurries back. 'What a red-letter day for this wretched coolie—to be mistaken for the brother of the Shōgun!' He thrashes him lustily for lagging behind. 'I shall beat him to death to allay your suspicions!' What agony for Benkei to strike his Lord! What agony for Togashi to watch Benkei striking his Lord! What delicious anguish for the audience to witness Togashi suffering for Benkei, Benkei suffering for his Lord, and his Lord suffering! Men of honour, all. Togashi lets the party through. Benkei prostrates himself before Yoshitsune and weeps. A song is heard: 'On mountain tops have we slept, in open fields, at the edge of the sea . . .' Togashi reappears and invites Benkei to drink with him. Benkei may not refuse . . .

They dip into their boxes of rice and fish and pickles, into their bottles. Tangerine peel flows round them. It lasts five hours. Emotions make you hungry. The connoisseurs nod lightly or smile tightly. Others roar applause. Women weep: 'Oh the pity!' Children stumble off to the *benjo*. An actor freezes into an extravagant pose, a stylized grimace. 'That's it!' they shout. 'That's what we've been waiting for!' They know a famous moment when they see one. Benkei entertains Togashi with stories, he dances, pretends to be drunk. 'Still thirsty?' someone jumps up and yells, 'I've got some *sake* left!' Wooden clappers signal the climax. (And wake a baby up: out comes an absent-minded breast.) To the sound of a drum, Benkei begins his triumphant exit, along the 'flower way' through the audience, taking his time, in ecstatic monstrous hops. An old man calls out the actor's name: 'You're every bit as good as your father was!' . . . A loudspeaker thanks them for coming, reminds them not to forget their belongings. Sighs are sighed, eyes dried, breasts tucked away, feet shod, legs stretched. Coats, umbrellas, babies, lunch-boxes are remembered. A great and good time has been had.

In his opinion, that was *real* culture, something always being talked about but, in his experience, terribly rare. He tried to remember his belongings.

Old legend

It happened that an ogre came among them
Attended by a female and a child of the race.
Child ogres are often winsome, as yet unspoilt;
The grown female defies classification.

The ogre belonged to the *sensei* clan
Which was reckoned to possess obscure powers—
At that period virtually every prefecture
Considered it politic to have its own ogre.

This ogre was a mighty eater of raw fish,
A ravager of inoffensive haikus,
Enemy of the *ofuro*, decrier of kimonos,
A constant complainer of cruelty to cats
(Yet he sought to gas their young in leaky ovens).
He uttered muddled speeches on Important Ogres
Of Past and Present, their Lives and Letters,
And issued proclamations concerning the delights
Of certain shady districts which the natives knew of
But did not wish to hear about.
A long nose, as the proverb says, is sure to poke.

The ogre was tireless, embarrassing the guardians
Of holy places with his would-be compliments.
Inebriated when all around were cold sober,
He offered to entertain the provincial gentry
With rude dances peculiar to his sort.
Among the festive, he sat beneath the cherry trees
Mumbling and shaking his head.
Often he approached virgins of good family, asking
If poverty had obliged their parents to sell them.
Rumour had it he was compiling a Chronicle
To explain the natives. This caused amusement
For he was totally ignorant of their metaphysics.

He meant well, they knew, he was an earnest ogre.
At last he departed. A small crowd escorted him
To the waterside. There he was heard to say:
'I am distressed by the weight of my sins—
Vouchsafe me a sutra.'
Or was it a suitcase? Legends are wrapped in mist;
Police records of the time are not reliable.

From the ashes

In new Berlin they open lesbian bars
Where new Brünnhildes smooth each other's hair,
And burn up Camels. They won't sell themselves
For all the fine tobacco in Virginia.

Whoever didn't do those dreadful things
(Ruins where decent people once read Rilke)
Have had some dreadful thing committed on them
(Darkened workshops of the late persuaders).

All ways it breeds a dour uncertain temper,
In prisoner and guard, in raped and rapist.
Also a weighty pride in stainless towers,
Bright highways, cheaper smokes, and opera.

We all adapt. Always there are bigger villains,
Other victims, somewhere. There are always
New theatres, new Rilkes, new persuasions.
Life must go on, with so much gone already.

Without end

>Close ties between old enemies,
>Swapping stories of lost sons or comrades
>At cocktail parties. Old soldiers
>In one another's arms.
>
>(Flags in the breeze; Spandau,
>Heerstrasse, Brandenburger Tor.)
>
>Chilly towards old neutrals—
>But how they detest old allies,
>Devious and unloving,
>Who cost them sons or comrades.

Sunny Siam

>The land was famous as a land of smiles
>('He who doesn't smile is ill,' they said),
>Its calmness ruffled by occasional coups
>So brief and cliquish as to pass unnoticed,
>Except a corpse or two uncovered later
>(Never deep down: digging was uphill work
>For people not exactly famous for it).
>Even the corpses wore a tortured grin.

Night Life

The busy bells of the trishaws
Chirp in the velvety night.
Here is a house of little fame,
Blue movies in black and white.

A woman attends the projector,
Ancient and prone to fail.
A baby is slung about her . . .
It's hard to make head or tail—

Surely the film is upside-down?
But no one dares to grouse.
Her husband is a policeman,
This is a private house.

All of a sudden the sound comes on,
The first sign of zest—
The baby is taking its pleasure
Sucking its mother's breast.

Desiring a beer

They sat outside in the garden, in a shady spot. You could just sit there quietly and have a drink. The madam brought them a girl. They didn't want a girl, they only wanted a cooling drink, they had driven a long way. 'Rather too old,' they said, being British and polite to the natives, 'if one may say so.' The madam returned with other girls. 'Still too old, really,' they shrugged disarmingly. Honour was at stake: at last the madam swept back in triumph. 'Cannot say too old, this one, eh?' They couldn't. Being British, they rose tremulously, thanked her curtly, left quickly.

Cf. Cocteau

Cocteau reckoned that opium 'chastens our ambitions';
Or else it lends a hand in making good the nicer ones.

One evening, two rowdies invaded the inferior shed
Where aching trishaw drivers smoke the dross of others:
A vulgar brawl, knives flashing, then one fell dead.
These were not patrons, quite obviously non-smokers.

'Lacking opium, a sinister room is merely sinister.'
But once the police had gone, deducing drink and passion,
All turned back to their pipes and their wishes—for
Calm, and clarity with calm, and absence of ambition.

The god faded

For a while, opium seemed the answer.
Peace, affection, and sharpened senses.

Was it poppycock? Opium smokers
Cadged fags on the side, forgot to wash,
Their flesh fell off, chests caved in,
Some would kill for the price of a pipe,
Senses sharpened to the coins in a pocket.

Also recommended are a full stomach,
Regular employment and even a good nature.

Pure accident

Roughed up by a dozen tipsy and excited policemen,
He found solace in sensing, between buffets, certain
Gradations of violence, even faint reluctances
(For whatever reason), and a pulling of punches.
He was comforted to note, between unfatal kicks,
A gun drawn but knocked by someone to the pavement.
All this arose from a pure and simple misjudgement—
In a time between Terrors, nothing to do with politics.

Even so

Even so. A house, a garden, near a river,
Little rivers in the garden. Cool of dusk.
Lamps flickering, mosquito coils, the scent of
Jasmine. A glass or two of Singha. Cigarettes
Like fireflies. Faint gleam of shirt and sarong.
Silent uproar of the crickets. Unweighty words,
Mostly unheard. Two silences. One peace.

But these words dim it. Are there things that
Can't be written, then? Allow a few. Repose.
It's gone, it's almost gone. And even so,
A house, a garden, water, silence, cool of dusk.

Sweet smell

At Bangkok airport, tickled by garlands of jasmine
From students and anonymous sympathizers,
He stands in a daze at the foot of the gangway
While smiling stewardesses interfere with him—
'Sorry, but we have to strip you of your flowers,
The scent's so heavy, the captain might pass out!'

They didn't last for long, those wreaths of laurel.
(He wonders what will happen to the booty,
Someone went to great pains in the weaving.)
Dragging up the ramp, his shirt smelling sweetly,
Puzzling how a poem could upset the ruling Marshal,
He hopes for a captain with a stronger stomach next time.

Singaporeans

Overseas Chinese are the easiest people to live among
They rob or assault or cheat only their fellow Chinese
And only when truly unavoidable or rewarding
They kidnap only the children of their fellow Chinese
(Unsure that foreigners care to recover their offspring)
They do not mind in the least what one may say
About the Malays and the Indians and the Eurasians
They do not much mind what one says about the Chinese
Since only the Chinese are to be taken seriously
Or more exactly those from the same village in China—
Indeed they are a most amenable people to live among
Not feeling obliged to have feelings about all and sundry.

In hot water

To debunk the borrowed robes of nationalism
He invented the term 'sarong culture' . . .
It was confused with the popular saying:

'When durians [*highly prized fruit*] come down
[*from tree*] sarongs go up [*supposedly referring
to* (a) *aphrodisiac properties of fruit, or* (b)
pawning of garments in order to purchase fruit]',

And hence taken to insinuate laxity both sexual
And financial among a section of the community.

The ruling party was not pleased with him at all.
'If he were Professor of Sociology, I could
Understand,' said the Prime Minister. 'But Literature—
What has that to do with real life?'

Interrupting the programme

An odd sensation, to enter a favourite bar
And hear oneself denounced on the radio.

(A change from queuing in some homely pub
To buy a pint, inaudible, ignored.)

The waiters were incurious. They had heard
So many scoldings, in so many tongues.

Moon-faced and beaming, to and fro they slid
With trays of drinks, not a drop slopped over—

Knowing that music was soon to follow,
Noble and wicked sentiments from Chinese opera.

Meat counter

'We're on your husband's side, mem!'
Thus the cheery young blades on the Meat Counter
In the supermarket. Chinese of course.
They chopped straight through a side of mutton
To show their sympathies.
He hoped it wouldn't come to that—
Frivolous as ever, drifting over to Wines and Spirits.
How noble the Scotch looked, how innocent the Gin!

Home

The old house was thought to be haunted.
The Chinese avoided it at nights—
During the occupation, or so some said,
The *kempeitai* used it for interrogations;
According to others, it served as a brothel
For Japanese officers.

There were fearful screams, and thuds;
There were grunts, and what sounded like
Rude guffaws and nervous giggles.

—A mass invasion of monkeys,
Bored with the nearby Botanic Gardens
And looking for new haunts.
They swarmed into the lavatory, tugged
At the chain, occupied the wooden seat.
They were a lively lot, they would go far!

One swung on to the power line,
Bit through the insulator, and sizzled.
They had discovered electricity.

And garden

What is more tonic, at times, than the jungle
Just held at arm's length, almost at your door?
Our respects, great Nature, man's best neighbour—
How you harbour chaos and yet preserve the law!
A grave macaque bent double like a gardener,
Raging bougainvillaea, thick-lipped frangipani,
A dewlapped lizard glides from branch to branch,
Or butterflies, those 'flowers returning to the tree',
Crickets and croakers, a snake or two in the grass,
Starling jostles bat till room is found for both,
And orioles of gold, and humming-birds like bees.
Ah the sweet smell of damp decay and sweaty growth!
Nothing can kill this world, you murmur as you ease
Yourself in evening cool, not even us, not wholly.
But don't forget, dilute your urine for the orchids.

Drinking with a Minister

After a ticking-off of dons by the Premier over dinner,
Drinking with the Minister of Health in the old Government House—
Everything was about to be forgiven, at least forgotten,
As the brandy went down and a new bottle was called for.
But the servants, a judicious mix of Chinese and Malays and Indians,
Turned sullen: they wanted to close up and go home.
'It is grown late,' they grumbled, adding a sardonic 'Master',
While rattling brass trays and switching lights off.
They had borne with the funny ways of the British,
From their own they expected more consideration.

Rest and recreation

>Why, it was almost Princeton! The GIs
>Scattered cautiously across the campus,
>Casing the idle buildings,
>Leaving notes in pigeon-holes:
>'Prof, I wanted to audit your lecture
>But we go back early Monday.'
>In Vietnam they'd forgotten about weekends.

The rival professor

Then a rival professor came on the scene. An American, sophisticated and *à la page*, an expert in linguistics. (So much more apropos than value-laden literature.) He was entrusted with a delicate official mission: the tutoring of young diplomats from Peking.

This put our man's nose out of joint. Not merely was he on the wrong side, but now it seemed there was a right side, and someone on it.

Abruptly the rival disappeared. His picture was turned to the wall, his name no longer heard in the halls of power. It was whispered that he had links with the CIA. (Unlike Wordsworth, Milton etc.)

The CIA was sedulous but prone to accident. (Admiring the PM's purity, it once offered him money. To the pure a cash offer is a bribe.) Yet how adroit it was in its choice of language.

Small honour

Things left undone, or done indifferently.
(As even he would come to see.)

Lectures that added blur to blur,
Pepped up with digs at the ministries;
Poems that started out in rage and pity
Then turned into harmless obsequies.
(To go no further.)

Counsel spooned out, the Wisdom of the West,
To keep his end up. Reasonably prudent,
And yet his great success: a criminal operation
Procured in secret for a negligent typist
Pregnant by an indigent student.
Such gratitude! She told the world.
(Later it seems he took against abortion.)

Courage enough, or carelessness, to get him into
Temperate trouble. A deportation order
On the cards. Phone tapped. Book banned.
Shocks that flesh and speech and print are heir to—
For whose good was it? His own small honour?

So long to learn, no time to be a teacher.
Better to give to beggars, no questions asked.
Then offered an OBE for 'services'—
His former pupils staffed the ministries.
(To go no further.)

You're not at home now

The ten commandments of expatriation
As drawn up in one's darker moments;
And later reduced to five, at the darkest
One being neither God nor Moses.

Confine yourself to value-free activities;
Advising the army, treating hernias in high places.
Otherwise keep to your ivory tower,
A contemptible address but unvisited
By the Special Branch.

Do not presume to complain of the heat
Merely because you see a native sweating.

Easy for you to like everybody;
Risky for anybody to like you.
Remember, the two old gods still rule:
Must-do and Make-do.

Judicious sex is permissible;
Sex has always been a substitute for politics.

Do good in secret,
Give to the Cancer Research Fund.
Cancer has always been a substitute for politics.

Your day will come.
City fathers will name a cul-de-sac after you.

Pom

Advance, Australia fair!
And so it did.

They didn't think he was a Pom.
(The Johore Professor of English

Had to come from Johore.)
They thought him 'a Malay of some sort'—

Albeit, like Australia, fair.
So he advanced.

In China

Lots of reproaches in China . . .
For not totting up the Hong Kong coins
In his pocket. And then
For frivolously declaring a trickle of cents.

For occupying a huge hotel room on his own.
And for proposing to share it.

For liking the cheongsam.
For suggesting it was the national dress.
For asking where it had gone.
('Gone to—ha—to decadent countries.')

For neglecting to press a cigarette
On the guide every five minutes.
For pressing a whole packet on him.

For scaring a group of small children
Into tears.
Then for making them giggle.

For being what was expected.
For being unexpected.

For not wanting to visit a collective farm.
For wanting to visit a university.

For mentioning *Sons and Lovers* in the staff room.
For declining to read a poem by Julian Bell
Out of Quiller-Couch's *Oxford Book*
On the grounds that it wasn't in it.

For being cultural instead of linguistic.
(Perhaps they could sense the Revolution.)

For enquiring after the Faculty of Law
('We do not have one! We do not need one!
We have no personal property and no divorce!').
For unlawfully upsetting the Dean.

For failing to laugh aloud at acrobats
Clowning around in US army uniforms.
For showing shock when a spectator hawked noisily,
There being a law against it.

For advising the guide to get some sleep.
For disrespecting the guide's solicitousness.

For ordering a second bottle of beer.
For starting the Opium Wars.
For offering his regrets.

For wishing to be friends.
For suspecting they might like to be friends.
For seeing that nobody could be friends.

He felt guilty if he enjoyed himself.
Enjoyment lay in feeling guilty.
Being scolded made him feel at home.

Did they feel at home, he would wonder,
When the Red Guards arrived soon after?

People's museum

In the museum hung a painting of tiny ships,
Flying huge Union Jacks and confronted by tall
And valorous peasants with cutlasses.

Another picture showed Commissioner Lin
Tendering ritual apology to the Spirit of the Sea
For dumping confiscated opium.

Before our time, the guide may have implied
By his strangled titter. The bemused Britisher
Grunted in muted apology—

Relieved that no one seemed to be denouncing anyone
At the moment; disturbed that so many seemed to be
Looking over their shoulders.

It was peace, for a time, in a place.
But the welding of nations is not so easy:
Old Mao was about to make waves in the Yangtse,
Without a word of apology.

Armies

As if so many princes weren't enough
They fashioned generals, often out of
Sergeants from colonial legions.

'Good iron isn't used for making nails'—
The people's wisdom sometimes fails.
Nobles are born, soldiers are made.

(He'd known a minor noble working in a
Filling station: 'Blame it on polygamy.'
Also a princess who taught philology.)

The women all looked like princesses.
Virgin corporals in battle-dresses,
They learnt the ways of the soldiery.

Dense with relationship, these sudden armies
(Be careful not to rape your sister!).
In extenso died the scattered families.

Still many princes left and many generals,
Removed to milder spots where rank still tells.

Off limits

Ponto-chō, heartland of the geisha, in Kyoto: a narrow street of wooden houses. One evening a jeep drew up at the open entrance to a little bar. Come out, ordered the US military police: Let's see you. To which our hero stoutly replied: You come in here. No, said the MP: It's off limits. (Not that the US Army had declared the bar off limits, but the bar had declared itself off limits to the US Army: there was this reciprocal arrangement.) Come out here, the MP repeated. Said he, insouciant: You can't arrest me, I'm English. A muttering under the breath, and away the jeep bounded, along the bank of the gently flowing Kamo.

But those dusty charming towns of long-gone Indo-China. Shady trees and hotels, peaceable slow-moving people, given to neatness (and to their neat children) and not-too-strenuous pleasures, graceful in their mild disgraces, a civilizing influence. Trishaw drivers chatting in French; now learning English, language of the future, a millennium of opulent fares. You can't arrest me, I speak English.

Process—how cruel it is. Nature abhors the vacuums we create, fills them with more like us. Filled women, cradles filled, streets filling up again. Ponto-chō stands where it did, a nest of singing girls. But gone are old Phnom Penh, Saigon. Come out; off limits. Such vanishings, such bans . . . like age—a kind of house arrest.

Wisps

>But these little ones are so very ignorant,
>It is hard to know where to begin with them.
>Mostly there is something to build on, to put to rights,
>Some special and meet desire—how we run to fulfil it!—
>Something to complete. Here there is everything to complete.
>All they wish for, impossibly, these wisps, is to be no more
>   than they were, no other.
>
>They block the roads. What realm are they subject to?—
>So skimped, unlicked, such blank unwritten pages.
>Folklore tells of a novice in hell, facing examinations,
>Who burns exemplary answers and swallows the ashes.
>They have swallowed ashes already.

What's to be done? While we ponder (we have to ponder?),
Let them be placed in—to borrow the word—in a camp.
For their own good, surely. (It has to be better than good.)
In a happy land, it goes without saying! Where sun shines softly,
    rain is gentle, peace drops slowly.
It isn't easy, best will in the world, in the best of worlds,
To provide for such guests, for these tiny ghosts.

The real thing

    Compassion was a man-made thing,
    A product of our affluent decency.

    These could not afford it. Something older
    Moved them: Stay alive, see to the children.

    Heads were kept down; or heads might fall.
    (Keep your foreign conscience tucked away.)

    In these, compassion was heroic, saintly;
    It had to be. Now and then you meet it—

    The real and purely natural thing.
    A privilege not to be forgotten.

The old days

    Whatever the creeds, the policies, the movements,
    Behind them straggle the ghosts of starving peasants.

    On this earth in the old days there used to be giants.
    No one reported the starving of peasants.

    For all his fears and flinching, at least he never
    Was much of a praiser of times past and over.

Lacuna

So little about the power of religion! And in regions renowned for powerful supramundane faiths. He might say, he saw little of it, it was obscured by the power of secular doctrine. Yet in one country, the student who believed he was Allah. In another, the Director of Textbooks (Min. of Ed.) who was quietly sure he was Buddha. Was he himself ever tempted to think he was Jesus? Not seriously. He recalled that Christ had been crucified. For a religion, mundane, all too mundane.

Spring clean

'If my devils were driven out, my angels too would receive a mild shock.' One takes the German's point, at once both poetic and prudent. But might perhaps the angels, mildly shocked, be less stand-offish then? If one cried, at most they murmured 'There, there', always having something better to do. They were rarely at home; one wasn't really at home with them. Immaterial seemed the word. Once or twice one may have entertained them unawares.

And afterwards, would the devils slink away, tails between their legs? Or merely strut more insolently? He could hear them jeering: 'Give a dog a good name, it'll still bite.' Devils were probably irredeemable.

Yet, oh to fall in love with a beautiful psychiatrist! With grey eyes, long neck, chaste white smock, motherly bosom. A wolf-man he! A ministering angel she! To take her away from it all. Or would she soon grow bored? (The *Standard Edition* of SF reads like a novel—like sf in fact.) Even though he told all. And it didn't do to tell women too much.

Or men, which most of them were. Flat-chested father figures. Or school doctors: 'More outdoor sport for you, young fellow!' Devils v. Angels at basket-ball? How politely Rilke declined the offer of 'a great clearing-up'. Heine would have emptied a chamber-pot over their dingy beards.

Better the devils you know . . . till the last dim beating of that other angel's wings.

The biographical mode

    Observing their subjects, biographers grow curious.
    Observing biographers, their subjects grow cautious.
    The bonfires of the latter make the former truly sore.
    Letters can prove a lot, but burning them proves more.

Naming names

A page in cipher—
But somehow sounding female.
Could be rhyming schemes
Possibly pet names of visiting typhoons
Perhaps the flowers in the Garden of Eden
Conceivably the peculiar terms in which a Thai princess addresses
    a Thai princess on the subject of Thai princesses.
(More likely the women he slept with.)

Watching one's words

    The word *homosexuality* implies the having of sex.
    Henceforth we shall use the expression *homoeroticism*.
    Undoubtedly there were men he was very close to.
    (Their names will occur no doubt in other contexts.)
    He can be said to have liked if not to have loved men.
    Hence we prefer to use the expression *homoeroticism*.

    The *homosexual* is licit now, and some say even gay.
    But what twisted tastes does *homoerotic* conceal?

Short thoughts

    Returning to your native shores, what is it
    Hits you between the eyes? The amount of dog shit
    In evidence. It may be heaven, it may be hell—
    One thing's certain: pets are used to eating well.

Pointing to Goethe's works, Freud identified evasion:
   'He used all that as a means of self-concealment.'
What a lot of self to conceal! What frailties unseen!—
A hundred and forty-three volumes in the Sophien edition.
Not much self in Freud, or else a lot of self-revealment:
In the end his *Gesammelte Werke* only comprised eighteen.

> Nature was best left
> To the specialists. It reminded him
> Of cross-country runs at school
> In wind and rain.
>
> Then there was Love. Severe girls
> At poetry readings chided him
> For not providing it.
> He left it to the specialists.
>
> From which it's easy to deduce
> A philistine and misanthrope.

They love us so much they could eat us.
Then get together to compare stomach pains.

But for them we would never have come here.
We must pray they never lose their appetite.

> Quite often heard to call on God—
> Though not expecting an answer.
>
> For who else could he call on—
> On some temporal lord and master?
>
> Angry with the one for being there;
> With the other for not, still angrier.

'You—the suave tempter promoting
An aperitif! Where's your dignity?
You, so—deep, so—distinguished!
You can't need the money.'

'You ninny—when God does voice-over
In commercials for nappies . . .
Dignity? Need? We move with the times,
The times are our business.'

It isn't bad literature we shall die of,
But good television: brilliant pictures that can
Only tell one story. Of what man does to man.
The camera can't lie all the time. Switch us off.

>	By nature pedantic
>	and deaf to ceremony
>	suspecting that agony
>	was rarely romantic.

>	The least one can do for one's family
>	Is dedicate one's books to them;
>	And the most one could do for them
>	Would be to swear off writing totally.

>	There used to be writers and readers
>	So readers would profit from writers
>	And writers would profit from readers
>	But the latter learnt more from the former
>	Than the former could learn from the latter
>	So the readers then became writers
>	And the writers no longer had readers.

He used to hate the laws. He broke as many
As he had the strength to. Or he bent them.

Now others have cast down the laws. He weeps
With rage to see them lie there, broken—

More laws than he had ever heard of,
Lying there, not bent, but broken.

In another time, another place they could have been friends. They had so much in common. Then and there it was not to be. We all have so much in common. In another place, another time we should be friends. Long ago and far away they would have been friends, in an earthly paradise, we should all have been friends. In a little time, under the earth, we shall be friends. We have so much in common.

> A founder member of the party of one—
> Then did he resign or was it expulsion?

>> At what age was innocence lost?
>> Not long after birth
>> It must have been. And then—
>> An eerie sense of arsy-versy
>> Or what Shakespeareans called
>> 'Inversion of the natural order'—
>> Seeming in pennyweights and scruples
>> Slowly to regain it. By what age?
>> Perilously near death it must be.

Posterity was always a great reader.
He would beg, borrow or steal books,
He would even buy them.
You could be sure to find Posterity
With his nose in a book.
(Except when listening to music
Or peering at paintings.)
He had excellent judgement too.
You could always put your faith in old Posterity.
We shall miss him.

That long disease, a Life: wind, diarrhoea, constipation, diarrhoea, earache, toothache, mumps, acne, black eye, common cold, headache, toothache, chilblains, pink-eye, cysts, prickly heat, chills, quinsy, arthritis, toothache, piles, headache, arthritis . . .

>> And he leaves to the members of his family
>> A Life of their own for each of them.
>> It is unseemly that guiltless bystanders
>> Be splashed by badly conducted vehicles.

Fear and knowledge

> Living next to an ancient primary school
>   TO THE GLORY OF GOD
> FOR THE BUILDING UP OF THE CHARACTER
>   IN HIS FEAR AND KNOWLEDGE—
> Maybe some of it would spill over, not only
> Toffee papers and finger paintings . . .
> Here they come. More mothers than children?
> No, just bigger. More cars than mothers?
> The children scatter from under their feet.
> Blonde dolls in peasant skirts and blouses,
> Pint-size scruffs in denims, parkas, plimsolls.
> Only the gleaming little black and brown boys
> In honest shoes, white shirts and proper trousers
> And a small trim oriental in (yes) a gymslip
> Look at all like British characters.
> They shall inherit top marks, for starters.

'Pipe-smoking fuddy-duddies the curse of publishing'

People fond of literature sometimes seek a career in publishing. This is a mistake, though not always fatal. A poet is creating in Reception: it is his right to be published, it is your duty to publish him. He is given a cup of tea. Then comes a person who has discovered the secret of immortality: it takes an eternity to tell.

Of course one needs living writers, but the dead are nicer. No risk of Mrs Woolf jumping off a bridge because her new book is full of misprints. Like the military expert intending to say: the loss of Vietnam will lead to a blood bath—and in the book it reads bloodless bath. Anguish. Horror. Commies at the printer's? Proofs are consulted—ah! spacing between words was uneven, and the author asked for *less* between blood and bath. Oh no, it isn't all bandying philosophy with Iris Murdoch over scampi and Chablis at Giovanni's.

He had his uses, an almost infallible judge. What he admired wouldn't sell and what he didn't would. Dark doubts assailed him. If you shrink from a tale of necrophilia, does it mean you are a crypto-necrophile? If stories of raped and murdered children fail to charm, that you harbour like urges? And another thing—if he could write books, why couldn't he write a decent blurb? Because blurbs need to be better written than books.

Ah well, if you want to go on liking literature, think of books as potatoes, sold in sacks. You like potatoes, don't you?

Tribute

>Caesar turns up each morning, asking breezily:
>   'And what are you offering this fine day?'
>Well, there's this thing—and that—and also
>   (you suppose) the other . . .
>The hours go by. Caesar has time to throw away.

>During the weekends he's away on his yacht,
>   or visiting Glyndebourne.
>God might take the opportunity to show his face.
>But would you recognize him after all this time?
>Best to put a few pennies by, just in case.

Things falling apart

An alarming sickness spread through the world, a plague which some called by an ancient name, individualism. It came between man and man, between man and wife, between man and master. Empires crumbled, Nations were discredited, the Parties fell apart, the Family broke up, Oddfellows grew odder by the day, the Acacia Avenue Tenants Association lasted but a week.

Hordes of individuals roamed the streets, like mad dogs, squealing like scalded cats, 'me, me!', doing their things, often quite nasty. (Had you avoided contagion? At times you had doubts.) Doubts were raging everywhere, by day and by night, and certitudes were not far behind. There was a great hunger and thirst, and little left to satisfy the needy. Some welcomed this course of events, declaring it a Sign of Life; often they retired to write their me-moirs. Others considered it uncivil, and blamed mistaken sages and false faith-healers or soul-doctors.

One alone there was who might stem this furious epidemic. But he had taken down his plate and vanished from the consulting rooms. A brainy though depressing German called Nietzsche insisted he was dead. Possibly he had been the first victim. That would be one of life's little ironies!

Parts of speech

>Forever rubbing shoulders with words,
>He couldn't always tell one part of speech
>From another, especially in the morning.
>
>There, inscribed on his desk in large letters,
>Was the verb REJECT.
>What a way to start the day!
>The grim faces of editors and women
>Stared up at him, ticket-collectors too.
>
>His eyes focused. There, incised on the pipe
>He reached for to start the day,
>Was the word REJECT in large letters.
>A noun, pronounced as reejekt.
>
>Never mind, he murmured, you're a good chap
>Whatever they say,
>And you're MADE IN GREAT BRITAIN.
>He walked off to work, cradling his pipe.

Where art thou?

He wrote a children's fantasy
Full of clean fun and unspoken morality,
For little Alice, Alice his ideal reader.

He ran to take her the very first copy.
But Alice was living with some feller,
Alice was on the Pill, Alice was crotchety.

He wrote it pretty swiftly,
It couldn't have been published faster,
But Alice was on the Pill already.

In the old days up at the varsity
They talked a lot about 'maturity'—
Could this be what they were after?

Oppressed poem

Who strained its stresses and reduced it
to mean things   Who starved its similes
and flogged its feet   Who seduced it
rhyme after rhyme and set the doggerels
on it   Who crushed it in cold ironies
Who sentenced its tender participles
to dangle and docked it of curlicues
Who hounded it into barren hyperboles
Who was the pronoun who martyred the muse—
Who shall be barred from polite anthologies.

Written off

Sad child, conceived without your willing,
Product of someone's careless living.

> Despised of men, if not rejected,
> By your sire's iniquities infected.
>
> Unlucky book, never to reach maturity,
> Whitely you lie there, staring through me.

Meeting writers

In unaccustomed comfort
Writers were travelling abroad to meet writers.

They met with experts on writers—
On the Writer in his Time
Or in his Place
On the Writer in Society
Or in Isolation . . .

> The writers did not meet with writers—
> Perhaps they were not nice people to meet
> Perhaps they did not care to meet writers
> Perhaps the writers had unwittingly passed them
>   travelling in comfort in the opposite direction?
> Or perhaps they were actually writing.

In any case writers convey an aberrant impression of writers.
Experts are always more trustworthy.

Old reviewer

> So long a reviewer, he could talk
> With a semblance of sense about most things,
> From belles lettres to birth control,
> From Tantra to the Toltecs.
>
> It was words did the thinking, he thought,
> And one word led to another.
> They thought of pleas to the tax man,
> Of poems and wills and testaments.

The whole of knowledge resides in words,
And words reside in the dictionary.
If someone stole his *OED* they stole his soul.
He slept with Roget underneath his pillow.

But he didn't sleep.
He thought of words like slumber, snooze, siesta,
Hibernation, aestivation, catnap, forty winks.

Cross words

Significant form: shapes of bottles, women, coffins.
Praxis: as long as it makes perfect.
Semiotics: more signs and portents!
Discourse: emetic to end a Chinese banquet?
Structuralism: you need a floor under your feet.
Hermeneutics: an interpreter is called for.
Deconstruction: books remaindered, books burned.
Tel Quel: but will they listen?
The Death of the Author: alas, an eternal verity.

Self-opinion

He might be thought to have thought himself
A middling poet because he came from a midland town;
Or provincial in that he had lived in Settsu Province
(Itself in the Kinki District, which was wasted on him);
Or Alexandrian, or (if hardly daily) Teutonic Knightly;
Or, having worked in the Black Country, demotic and smutty;
Or lionlike in coming into Singapore (and out like a lamb);
Or capital since (with the help of it) he moved to London,
But south of the river, so a simple suburban soul,
He might be thought to have thought himself.

Ambitions

What he would have wished . . . a dozen surreptitious ambitions . . .

To have grown the first potato

To have invented one or more of the following: anaesthetics, the flush toilet, the Golden Mean

To have uninvented no end of things, including the end

Alternatively, to have composed requiem masses noisy enough to waken the dead

To have written like George Herbert (and lived longer)

Or to have written (parts of) the Bible

To have founded the NSPCC, the RSPCA

To have been a hero (without actually killing anybody)

To have been boss (short of bossing people about)

To have been Pinkerton and changed the story of Chōcho-san

To have possessed this man's art and that man's scope (without needing to be either of them)

To have been himself, with certain improvements.

Great chain

> The emissaries of Venus
> Run riot among the guardians of culture
> (Male and female are they also)

> Like a tracer, the twisting streptococcus
> Plots the intellectual luminaries
> Like a string of bulbs on a Christmas tree
>
> Like a radioactive isotope
> The holy spirochaete
> Links the media in spirals of loving
>
> Who reviewed whose book and the wages thereof—
> For the eye is not satisfied with seeing
> Nor the ear with hearing
>
> And would there be peace now
> While the high ones lick their wounds
> Would the thunder of brains desist
> While physicians attend the lower regions
>
> No, for the clever are not like us
> Pain is a spur that the clear spirit doth raise
> To lead laborious nights and days—
> Here come more doses of great fame.

Sadder, the massacre of an innocent, a symbolic event. Poul was an artless young Dane, melancholy, sensitive, shy. Never had he heard those bold words: 'On our sleeves we wear our sexes,/Our diseases, unashamed.' In Krung Thep (we prefer the official name on this occasion) they always asked him to their parties. Too shy to speak to anyone, hardly drinking—but he liked to come, he said, he would just stand and look and listen. Would he had done no more.

Poul disappeared. Had he gone home on leave, too modest to mention the fact? They discovered him in hospital, a case of advanced syphilis. Resolving to wean himself away from shyness, he had gone with a woman. Too shy, too modest, to look elsewhere, he went with a prostitute. Too shy to go to a doctor. It was his first woman. His first job too. Shocked, alarmed, the East Asiatic Company sacked him. Such can be the wages of shyness, sensitiveness, sex.

Fathers

Visiting Belfast, he would hardly have boasted
Of how his father boasted of a Fenian father.
Remembering rather that many years before
His father was taxed with the small explosives
Stuffed into pillar-boxes—
The boxes that it was his father's job to empty.

Small cultural objects

'America, you have it better,' Goethe observed
Plaintively. Certainly better than old Poland.
For instance, the carefully faded elegance
Of the Hotel Europejski, once the pride of Europe,
Its marble stairs redeemed by a ragged carpet.
In the foyer, behind large slices of plate glass,
Small cultural objects cower, for sale or not,
And a native poet confides to a foreign one
That three poems have been banned from his new book—
Should he withdraw it? Or is half a loaf better . . .?
An enormous radio is transfixed to one transmitter:
All the better to be heard. Outside the chamber
Sits a deadpan chambermaid. Doubtfully for sale.
Servants are not servants, they are civil servants,
They wish to meet your wishes, which had best be modest.
Back in the far-flung shadows of the restaurant
A cluster of rustic hussies try to remember
What cocottes were like. The Palm Court players
Explode in a waggish folksong, and they stop trying.
Waiters whisper hungrily, something about dollars.
Are they provocative agents, like the hussies?
Here are the grand menus still (stained with bortsch),
But not the dishes. Half a loaf . . .

The elusive sage

The so-called 'Sayings' occupy a thick exercise book, but for present purposes a handful of examples will suffice. To begin with, there is an element of mystery attaching to the urbane Chinaman whom he claimed as a lifelong friend and mentor. Sinologists are of the opinion that Tao Tschung Yu probably lived (if at all) in the eighteenth century. Many of the dicta support this view.

> 'The Emperor speaks proudly of the young men he sent to their deaths; in the palace window a candle weeps for a straying prince.'
>
> 'Who sorrows for the long nails of the mandarin? Only his wife with the bound feet.'
>
> '"Too thin-skinned for words!" complained the cow as she rolled over on her friend the butterfly.'
>
> 'Poetry teaches the correct names of birds, beasts, herbs, trees, persons and places. Do you ask for more?'
>
> 'He who makes an offering as the temple burns down is an idiot; or conceivably a superior person, psychologically speaking.'

But was there psychology in the eighteenth century? Possibly the Chinese invented it. Many of the 'Sayings' strike us as curiously modern in tone.

> 'The secret of an orderly and stable regime: writers should be obscene and not heard.'
>
> 'The Minister for Culture wants one urgently? Place some agar-agar in a saucer . . .'
>
>> 'Politics has overrun our beds—the last border!
>> It lusts for us, the thing we lusted after.'
>
> 'People may be divided into two classes: those who think that if weapons of war abound they will not be used, and those who think the contrary. A third and more numerous class prefer not to think.'

> 'With so much violence around we shall soon see a war.
> With so much violence around we do not need a war.
> With so much violence around we have a war.'

> 'The work of the devil who rules over printing-houses: *world* supplanted by *word*.'

> > 'They juggle with some great abstraction,
> > And study the small print.
> > Their love for man is beyond calculation,
> > They keep a numbered account.'

> 'Like mothers, critics wish you wholly other than you are. But they do not feed and clothe you.'

> 'A dream: the awarding of bursaries to non-writers to enable them to continue not-writing.'

> > 'Line after line incised: that's character!
> > Then the knife slips. Back to the maker.'

> 'That great guy Paterfamilias—saint? philosopher? dreamer? brute beast? Whichever, let us honour him.'

The sage can occasionally sink into facetiousness, e.g.

> 'Who are the most illustrious of the people of Ch'in? Fu Manchu and Confucius He Say',

and although the following aphorism is enigmatic in the Eastern fashion,

> 'As the noun declines, so the verb fails to conjugate',

scholars point out that Chinese is uninflected, and hence the language does not lend itself to the putative metaphor.

Did he really have a friend called Tao Tschung Yu? Or was this a man of straw, a stalking-horse, a mouthpiece for dubious wisdom? The final entry may conceal a clue.

> > 'Avoid him who forgives himself everything.
> > Avoid him who forgives himself nothing.
> > Cherish him who even forgives this moralizing.'

Sleeper, awake

A poet told of enjoying in a dream the ugliest and
dirtiest girl imaginable, when for the same price
he could have had the most beautiful princess . . .
In our dreams we are often surprisingly modest.

In his dreams he was customarily fairly modest,
apart from catching the youthful Princess Elizabeth
when she tumbled from what seemed a circus elephant,
and once composing a late-Jamesian novella . . .

So he knew he was sick, when instead of missed trains
or friends, or missing lecture notes or trousers,
he dreamt of prodigious signals flashing in the sky,
and singular emotions, and the world's end imminent.

Even then not much happened. At the eleventh hour
he woke in a sweat, and found it well past midnight.

Fears

It couldn't always end in clownish despair.
There must have been real depths somewhere,
And even relative heights. He must have won
Something. Sometimes have heard *koi* rightly.
More fearful he was than an ancient Briton—
Wood touched, fingers crossed, lips sealed tightly.
Evil had eyes. Hate lay in wait for love.
Fatima's hand fitted him like an old glove.

What you love, never like too much. That which is given can be taken away. He jests at crossed fingers who never felt a loss. Lonely the glove whose mate is mislaid. Idle talk can cost the lives you love. Drowning in depths, falling from heights—man is a reed that barely dares to think . . . These are our homeopathic bromides, not to be scoffed at.

As for 'hearing rightly': maybe, but a true story lasts longer than a sham lay.

Love poem

'I did not promise you a rose-garden
I did not mention an epiphany
(Yet you are the wisest of readers)

Nor, although pushed, has language fallen
Never did sin come into the word
(Even if innocence has long departed)

I merely offer whips and guiles
Cods and wrecks and wreaths and biles
Roughage by which our minds are moved
Fallen or sitting or rising up

Gladly would I give you a rose-garden
If life were a bed or I had green fingers
If I were Christ, I would think quite seriously
Of arranging a grand epiphany.'

(Though described in a note as 'about the fall of language etc.', this must surely be one of his rare love poems, and it is printed here as such.)

Like mother . . .

Click, click! She adding to the available stock of cardigans
—For his birthday she might give him a good grey pullover—
And he, though less equably, to the stock of phrases.
Busy hands, each carefully tolerant of the other's fashions
—Once he dedicated a nervous book of verses to her—
While wondering who would want the words, who wear the woollies.

Dropping names

A pusillanimous view!—that you cannot decently write about the living unless they are unrecognizable, or illiterate, or live at a safe distance and will never know, or are as near death as makes no difference. (Or, very occasionally, are indubitable and unmitigated scoundrels; or established saints.) Or of course—as his mother became—they are virtually blind. (What she most missed was knitting, and television, specifically the advertisements, which she found more consistently agreeable than the rest of it.) Since self-denial of this kind affects the most interesting things in life, such *mauvaise honte* is a great shame.

Around the Styx

      Nothing of a Berryman,
      Closer to a ferryman—

      Hanging around the Styx
      For evidence of dirty tricks

      Where come the dead and dying
      With their unregarded crying

      (Charon describes his passengers
      As mostly innocent bystanders)—

      Or, piling man-made miseries
      On natural calamities,

      Typhoon and earthquake, fire and flood,
      Costly ideologies, cheap flesh and blood,

      More like Old Moore's Almanac
      Engrossed in universal wrack.

Untold

Memory, our nurse and friend and patron—
Until by freakish maladministration

She summons up those ghosts, both new and old,
Kind and unkind, whose names we leave untold

For fear of giving flesh and blood and bone
To what we then must own to or disown.

No end in sight. They spread from coast to coast.
Once heaven or hell was where we sent each ghost.

Behind, in front

He had long felt himself implicated in that
'terrible aboriginal calamity' noted by Newman;
whatever its purposes, creation was out of joint.

Curious then that with the years there grew a sense
that this calamity was yet to come, this huge mishap
of purpose, in some ever-approaching future;

no abstract primeval perversion incurred out there
among the infant planets, but not so far from here
and now, the putting out of joint of him.

Actuarial

They're all agog to handle your cash
Even though you have none—
But though you still have one
No one wants to underwrite your life;

Only your roof, your bed, your desk,
That faithful if unsteady clan.
Approach the Capital Builder Plan
And you'll be shown the door.

(The cult of youth was founded
Soon after he turned forty.
Soon after he has ceased to be
They'll introduce eternal life.)

When things seem rather safer
Than they often were, it's sad
That you've become so plainly bad
A risk. Well then, be bad, take risks.

Auguries of sinfulness

Be bad? Less easily done than said
(and if done, not to be read).
What is the wickedness reserved for age—
to put the National Health in a rage?
Arson's as out of place as Sodom;
the poor-box electronically secured;
language already well manured.
When the blood runs sluggishly
conscience can catch up quickly.
Seriousness seeps in,
like syphilis (*v. supra*) . . .
One is old enough to be anybody's grandpa.
(Maybe they do it different ways
these days.)
After four glasses, the love-seat calls;
after eight, sleep falls.
(Coffee may inhibit drowsing,
but isn't positively arousing.)
Take risks? Certainly, Miss Risks,
where would you like to go?
Burning at both ends ah! and oh!
it gives a lovely light,
but will the game prove worth the candle,
let alone burnt fingers and a scandal?
Will be? Won't be? On the whole
you might find more delight
in questing for the soul—
that haggard sexy heroine,
the eternal feminine.

Where can it be?

The seat of the soul. Hardly in the purlieus given over to pleasure. Possibilities were struck off one by one: the nose, blocked with grime, prone to offend its owner; the eye, ever looking for offence; the all-too-busy hand. (How can it be, as a modern sage declares, that the body is the best picture of the soul?) Until only the zones of pain remained. Should one suffer solely from gout, would the soul therefore reside in the big toe? Well then, the breast, or the heart? Once highly desirable neighbourhoods but now overrun by lawless pangs and migrant uneases. Or the head, the capital itself? Smoggy, littered with grubby worries, addicted to aspirin and other drugs. Or, as some holy man surmised, the pineal gland? A 'pea-sized organ in the brain'. Roominess was immaterial, but even so, not a very grand address. If no suitable seat could be located—would this suggest there was no soul?

Secret drawer

> The secret drawer locked against burglary,
> What did it conceal? Nothing so innocent
> As a prescription for glasses, a rosary.
> No, but a bundle of letters his agent
> Had written him over the years, doggedly
> Urging him, with little sign of resentment,
> To tackle something major—a biography.

Instruments to plague

> As time went by, they found you out,
> In failing sight, strange warts and gout—
>
> He traced each single dolour to its dim
> Unvirtuous place, now costing him . . .
>
> Though still aware of those, the utterly
> Innocent, struck down already, finally.
>
> That there was justice in the world he saw;
> And of injustice rather more.

The evil days

While the years draw nigh when the clattering typewriter is a burden, likewise a parcel of books from the postman, and he shall say, I have no pleasure in them; for much study is a weariness of the flesh;

Also when the cistern shall break, and the overflow be loosed like a fountain; when the lights are darkened, and the windows need cleaning,

And the keeper of the house shall tremble at rate bills, and be afraid of prices which are high; and almonds are too much for the grinders, and beers shall be out of the way;

Yet desire may not utterly fail, and he shall rise up at the sight of a bird, when the singsong girls are brought low;

In the day when he seeks out acceptable words; when the editors are broken at their desks, and the sound of the publishers shall cease because they are few.

Then shall the dust return to the earth as it was, and the spirit also, whether it be good or evil, shall look for its place.

Waiting to be served

>The boss can't spit in every single dish,
>Or leave the bones in every fillet of fish
>Or (it would cost the earth in bananas)
>A banana skin for each of his customers?—
>
>Except he is weary of numbering
>The fall of sparrows and farthings.
>We shall all of us get the same dessert.
>No more dying alone, or lying in state—
>Form up in queues and wait.
>
>. . .
>
>Don't play with me, Lord. I don't want
>To be understood. Nor to hear about
>Extenuating circumstances.

Do we all get in by the back door, then?
Please—it's neither the time nor the place
For theology. And I never asked to be saved
By others dying on a battlefield or cross.

Better to be punished in hell than patronized
In heaven. I can understand your critics.
Rather that other court of justice,
Handing down plain verdicts—

Where sin can enter by the big front door
And feel at home.
We want to feel at home, Lord.

(God as a spiteful chef? As a benign bumbler? Some people are hard to please! It seems that what used to be known as 'religious mania' is still found in a warped form among the irreligious.)

Last things

You need to be a Catholic to stand that sacrament
Going on around you. Otherwise, passionless faces
Of nurses, and someone you vaguely know, saying
The worst is over and you'll soon feel much better.
Probably you will. It can't hurt to leave with the
Sound of truth in your ears. Unless you guess at
Rewards or reprisals ahead, and doubt that better's
The word for you. In which case the last rites are
Not so bad. (Extreme unction has helped you before.)
Nor, come to that, is the needle, that blunt visa.
At least go in peace, since you must go, in peace.

Peace, here, being primarily a physical affair. This is the body's hour. The soul, wherever situated at the moment, has all eternity before it. It's the flesh that wants to be put decently to sleep. The spirit, with a care for residual dignity, will not disapprove. May even concur. They were closer, over the years, than was often supposed.

Ever-rolling

Though much is uttered at many assemblies
We fail to hear the holy voice of Clio.

History perhaps has ceased to make herself;
More than enough, she thinks, to last for ever.

Like brilliant fog, the dead, the famous ones,
Eat up our oxygen, and nod politely—

At every man a king, king for a day, then
Lost in a common last unfinished phrase . . .

Those quivering mounds of flesh and tatters—
Matter for some new Gibbon or Macaulay?

A score or more of leading international poets
Mount a platform. The earth declines to move.

Some thought they saw the shade of Thomas Mann
Linger by a news-stand, a pained look on its face:

'After me a deluge. Merely a flood.'
Perturbed of course, and yet a trifle smug.

While ever-rolling streets bear shades and
Solids, agents, patients, all away,

And still we sit in chains, where film
Unwinds its strings of undistinguished horrors.

If all the innocents, the slaughtered commons,
Swollen children, if only they could rise

In one great cloud to heaven, at least
The guiltless, could go and leave us free,

Leave us a yet unwritten page. A virgin year,
And this tired soil lie fallow. What might follow?

We thought we prized the past, its noble gifts,
White elephants that ate us out of heart and home.

If on the moon mankind could lose its memories;
On some fiery star our brains be wiped quite clean.

Until which time we make our unfresh starts
And share our instant chronicles. It's your turn now.

INDEX

Lives of the poets	1
Prefaces	1
Great debate	3
Explanations	3
Absit omen	4
Catch	4
Stones	4
Appearances	5
Innocence	6
Question of degree	6
Circumstance	7
Animal lover	7
First impressions	8
Going cheap	9
Do not litter	9
No show	9
Entertaining women	10
A child's view	10
Refinements	11
Samurai	11
At the theatre	12
Old legend	13
From the ashes	14
Without end	15
Sunny Siam	15
Night life	16
Desiring a beer	16
Cf. Cocteau	17
The god faded	17
Pure accident	17
Even so	18
Sweet smell	18
Singaporeans	19
In hot water	19
Interrupting the programme	20
Meat counter	20
Home	20
And garden	21
Drinking with a Minister	22
Rest and recreation	22
The rival professor	22

Small honour	23
You're not at home now	23
Pom	24
In China	25
People's museum	26
Armies	27
Off limits	28
Wisps	28
The real thing	29
The old days	29
Lacuna	30
Spring clean	30
The biographical mode	31
Naming names	31
Watching one's words	31
Short thoughts	31
Fear and knowledge	35
'Pipe-smoking fuddy-duddies the curse of publishing'	35
Tribute	36
Things falling apart	36
Parts of speech	37
Where art thou?	38
Oppressed poem	38
Written off	38
Meeting writers	39
Old reviewer	39
Cross words	40
Self-opinion	40
Ambitions	41
Great chain	41
Fathers	43
Small cultural objects	43
The elusive sage	44
Sleeper, awake	46
Fears	46
Love poem	47
Like mother . . .	47
Dropping names	48
Around the Styx	48
Untold	49
Behind, in front	49
Actuarial	49
Auguries of sinfulness	50
Where can it be?	51
Secret drawer	51

Instruments to plague	51
The evil days	52
Waiting to be served	52
Last things	53
Ever-rolling	54